ALL ABOARD AMERICA

Washington Monument

ABDO
Publishing Company

A Buddy Book
by
Julie Murray

VISIT US AT
www.abdopub.com

Published by ABDO Publishing Company, 4940 Viking Drive, Edina, Minnesota 55435.

Printed in the United States.

Edited by: Christy DeVillier
Contributing Editors: Michael P. Goecke, Sarah Tieck
Graphic Design: Deborah Coldiron
Image Research: Deborah Coldiron
Photographs: Corbis, Hulton Archives, Library of Congress

Library of Congress Cataloging-in-Publication Data

Murray, Julie, 1969-
  Washington Monument / Julie Murray.
      p. cm. — (All aboard America)
  Includes index.
  Summary: Discusses the construction, history, and current status of the Washington, D.C., monument honoring George Washington and some of the famous events that have taken place around its base.
  ISBN 1-59197-509-3
     1. Washington Monument (Washington, D.C.)—Juvenile literature. 2. Washington Monument (Washington, D.C.)—History—Juvenile literature. 3. Washington, George, 1732-1799—Juvenile literature. 4. Washington (D.C.)—Buildings, structures, etc.—Juvenile literature. [1. Washington Monument (Washington, D.C.) 2. National monuments.] I. Title.

F203.4.W3M87 2004
975.3—dc21

2003051922

Table of Contents

The Washington Monument is a **memorial** to George Washington. He was the first president of the United States. The Washington Monument also stands for American freedom.

The Washington Monument is more than 555 feet (169 m) tall. It stands in Washington, D.C. Washington, D.C., is the capital city of the United States. A capital city is where government leaders meet.

The Washington Monument
(WAH-shing-tun MAHN-yuh-munt)

CHAPTER 2

George Washington

George Washington was born on February 22, 1732, in Virginia. He led American soldiers in the Revolutionary War. With his help, America became a new nation.

Geroge Washington

Americans know George Washington as the "Father of His Country." He helped to build a government that has lasted until today. Washington served as president from 1789 to 1797.

On December 14, 1799, George Washington died. That year, Congress agreed to build a tomb for him. It would be inside the nation's Capitol building. But Washington's family did not want to use it.

In 1833, John Marshall and James Madison formed the Washington National Monument Society. The Society planned to build a large monument honoring George Washington. Marshall became the Society's first president.

The Society began raising money for the monument. In 1836, they held a contest to find a designer. Robert Mills's design won. Mills's design was a round building with columns. A tall **obelisk** would stand in its center. The Society believed the obelisk alone would work as a monument.

Robert Mills's design of the Washington Monument.

Thousands of people gathered on July 4, 1848. They celebrated the Washington Monument's first day of construction. President James K. Polk attended the celebration.

People celebrated the Washington Monument's first day of construction.

Workers began building the **foundation**. They used marble from Maryland to build the **obelisk**. Construction stopped in 1853. The Society had run out of money.

The Society asked people to give **commemorative** stones. The stones were built into the monument. People from around the world sent stones. One came from Pope Pius IX in Rome, Italy. But the Pope's stone was never used. Someone had stolen it.

In 1854, construction stopped again. At that time, the Washington Monument was only 152 feet (46 m) tall.

The unfinished Washington Monument in 1860.

In 1854, another group took over the Washington National Monument Society. They were called the Know-Nothing Party. The Know-Nothings were secretive. They often said they "knew nothing" when people asked about their group.

The Know-Nothings were against the spread of the Catholic religion. They had not wanted the Pope's stone to be part of the Washington Monument. Had the Know-Nothings stolen the Pope's stone? No one is sure.

The Know-Nothings were in charge of the Washington Monument for three years. In 1858, people from the first Society took over. But there was no money to begin construction. Nobody worked on the monument for many years.

In 1876, the United States government took over the Washington Monument project. The U.S. Army Corps of Engineers was in charge. Lieutenant Colonel Thomas Casey was the leader of the project.

The Corps made the **foundation** much bigger. They also built a special steam-powered elevator. They used it to lift marble blocks. They used the same kind of marble. But it did not exactly match the rest of the **obelisk**.

On December 6, 1884, a special stone was placed on the **obelisk**. It was pointed at the top. It weighed about 3,300 pounds (1,497 kg). A smaller **aluminum** cap went on top.

Workers placed an aluminum cap on the top of the Washington Monument.

A celebration took place on February 21, 1885. It was the day before George Washington's birthday. People gathered to honor the finished Washington Monument. President Chester Arthur dedicated it "to the immortal name and memory of George Washington."

President Chester Arthur

Visitors were allowed inside the monument on October 9, 1888. A stairway led to an observation deck. People could ride an elevator to the top, too. The ride took about 10 minutes.

A view from the Washington Monument's observation deck.

Many people visit the Washington Monument each year. There is an Interpretive Center on the ground floor. It has special displays about George Washington and the monument.

Views from the top of the Washington Monument are great. Visitors can see the Lincoln **Memorial** and the Reflecting Pool. They can see the White House and the United States Capitol, too.

The Lincoln Memorial, the Washington Monument, and the Capitol.

People gather at the Washington Monument for events, too. Important speeches and **rallies** happen there.

People go to the Washington Monument every year on the Fourth of July. This American holiday is Independence Day. Americans watch fireworks and celebrate their country's birthday.

Detour ⬇

Did You Know?

- It cost $1,187,710 to build the Washington Monument.
- The Washington Monument's base is about 55 feet (17 m) wide.
- The Washington Monument weighs more than 90,000 tons (81,647 t).
- The stairway to the top of the Washington Monument has about 890 steps.
- Inside the Washington Monument are 193 commemorative stones.

Important Words

aluminum (uh-LOO-muh-num) a lightweight metal that does not rust.

commemorative (kuh-MEM-ruh-tiv) describes something that was made in memory of a person, place, or event.

foundation (fown-DAY-shun) the base that helps to support a building or structure.

memorial (muh-MOR-ee-ul) something that reminds people of a special person or event.

obelisk (AH-buh-lisk) a tall pillar with four sides and a pointed top.

rally (RAL-ee) a large meeting to discuss something important.

Web Sites

Would you like to learn more about the Washington Monument?

Please visit ABDO Publishing Company on the information superhighway to find Web site links about the Washington Monument. These links are routinely monitored and updated to provide the most current information available.

www.abdopub.com

Index